FIGHTING FORCES
ON LAND

M1A1 ABRAMS
TANK

DAVID BAKER

Rourke

Publishing LLC
Vero Beach, Florida 32964

www.rourkepublishing.com

PHOTO CREDITS: All photos courtesy United States Department of Defense, United States Department of the Army, General Dynamics (Land Systems Division)

Title page: *Bargaining with the locals. This tank crew talk with Iraqi civilians in a splendid view of the Abrams with turret reversed 180 degrees to face back across the raised engine covers.*

Editor: Robert Stengard-Olliges

Library of Congress Cataloging-in-Publication Data

Baker, David, 1944-
  M1A1 Abrams tank / David Baker.
    p. cm. -- (Fighting forces on land)
  "Further Reading/Websites."
  Includes bibliographical references and index.
  ISBN 1-60044-248-X (alk. paper)
  1. M1 (Tank)--Juvenile literature. I. Title. II. Series.
  UG446.5.B2345 2007
  623.7'4752--dc22
                                        2006011682

Printed in the USA

CG/CG

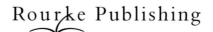

Rourke Publishing

www.rourkepublishing.com – sales@rourkepublishing.com
Post Office Box 3328, Vero Beach, FL 32964

TABLE OF CONTENTS

CHAPTER ONE

★ TANKS IN BATTLE

The tanks that first appeared during World War I (1914-1918) were simple, big, and very slow. Today, the modern tank is an efficient fighting machine capable of accurately hitting targets in day or night conditions, in clear weather or in fog, at rest or at full speed. There is no other tracked fighting vehicle that carries so big a punch and can put such devastating **firepower** on enemy positions. Capable of taking on some of the most heavily defended positions; they can punch through a **blizzard** of enemy fire and survive to destroy their targets.

▲

Modern tanks are capable of great destructive power and carry advanced technology to help navigate and aim their guns with great accuracy. This M1A1 is seen at speed across the desert.

▲

This side view of an Abrams shows the clean, low profile lines of the hull and turret. Gyroscopically controlled, the gun can maintain full lock on selected targets. Note the space between first and second road wheels.

Today tanks use **satellite** information to accurately bring their guns to bear on target. They gather information from **infrared** and optical sensors fed to computerized systems controlling the tank's **navigation** and targeting systems. Advanced and exotic technology is at the heart of the modern Main Battle Tank (**MBT**), the heavy punch for assault forces with mobile hitting power flexible enough to fill a wide range of duties on the battlefield or in dense urban warfare.

PLANNING FOR THE ABRAMS

Tank design has changed greatly over the decades, evolving from cumbersome and slow armored vehicles to powerful tracked fighting machines. By the end of World War II (1939-1945) the Germans had tanks weighing up to 50 tons with guns capable of penetrating armor at a range of more than one mile while moving across rough **terrain** at speeds of up to 30mph. By the end of the war in 1945, tanks had become one of the most important weapons on the battlefield.

▲

General maintenance and repair is much easier with the Abrams compared to earlier tanks. Time spent in repair keeps the tank out of action.

▲

The track layout rides on two sprocket wheels and seven road wheels and is protected with armor plates that can be removed or exchanged when damaged.

With new threats from aggressive Soviet forces, the world plunged into the Cold War where the United States and its allies faced a communist threat masterminded from Moscow, the capital of Russia. The United States needed a tank capable of outgunning the Soviets and her allies in Eastern Europe that threatened to lunge across the north German plains and attack the West. Named after General Creighton W. Abrams, former Army Chief of Staff and commander of the 37th Armored Battalion, the M1A1 Abrams was the answer.

▲

Keeping the design as low in profile as possible is a compromise between ground clearance and height as seen here in this view of an Abrams on patrol. The gunner's 7.62 mm M240 machine gun can be seen to the right of the turret top.

▲

Stocking up. A soldier transfers a shell into the stowage bins inside the Abrams' gun turret.

Its purpose was to form the backbone of US armored formations, to engage the enemy in any weather, day or night, close with and destroy any opposing fighting vehicle in the world and do so with protection for its crew in any conceivable environment.

▲

The front of the M1A1 Abrams main battle tank is designed to provide minimum area and to deflect incoming shells. Special armor protects the crew from every known type of ammunition.

▲

In designing the Abrams, engineers shaved any excess surface area to give enemy gunners the leanest target, as displayed here on the Marine Corps Abrams. Note how the hull is shaped down forward of the engine compartment.

CHAPTER THREE

DESIGN FOR A SOVIET THREAT

The decision to build a new tank was taken in 1972 and the first **prototypes** were being tested by 1976. Within four years the M1 was ready for service. In February 1980 the first of the new Abrams series were being delivered to US Army **battalions** for evaluation. By 1985 the new model, the M1A1 was delivered to army units and in the early 1990s the M1A2 was in production.

▲

M1A1 Abrams share parking space as their crews prepare for a reconnaissance mission.

▲

Tank turrets are designed not only to house the gun crew and commander, but also to carry extra equipment.

With a crew of just four men the M1 has an **angular** appearance. The driver sits in the center just forward of the turret and the other three crewmen are in the main body of the revolving turret. The gunner and the commander sit one behind each other on the right side of the turret and the loader sits on the left. The driver has three **periscope** vision blocks for forward and sideways view but the center block can be replaced with a low-light periscope.

Built by General Dynamics Land Systems Division, the 60 ton M1A2 Abrams is powered by a 1,300 hp Lycoming Textron gas turbine engine located in the rear and coupled to a transmission system providing four forward and two reverse gears. The M1A2 can travel at speeds of up to 42 mph and the tank has a normal range of about 250 miles on 500 gallons of fuel, depending on terrain and fighting conditions.

Ploughing the sand. An Abrams with a forward plough clearing obstacles. Fuel consumption is high when carrying out these duties, albeit a necessary part of opening a path for less heavily armored vehicles.

▲

The commander's 0.5 caliber M2 machine gun is seen here fronted by a spare road wheel attached to the side of the turret.

The main **armament** is a 120 mm smooth bore cannon with 40 rounds and a kill range of 13,000 feet. Two 7.62 M240 machine guns are carried in addition to the commander's 0.50 caliber M2 machine gun, all three of which have a total 12,000 rounds of ammunition.

The Abrams tank has generous storage space around and on top of the flat and faceted gun turret. Note the commander's 0.5 caliber machine gun.

WARS AGAINST IRAQ

For ten years, after the Abrams was delivered to the US Army it remained untried in the real world of war and conflict. At the end of the Cold War, it seemed the Abrams was now without a role, its original purpose of fighting Soviet forces now redundant, but there would soon be a new demand placed upon the tank and in a very different part of the world.

▲

Front wheel guards are easily damaged. Tanks rarely stay in their factory condition and soon bear the scars and marks of battle, each one personalized by the crew.

▲

CHAPTER FOUR

Patrolling a remote region of Iraq, communication is an essential part of supressing terrorist and guerrilla activity. Note the spare wheel on the turret top to the right.

▲

An Abrams gunner discharges a 155 mm round against a target in Iraq.

When Iraq invaded the independent state of Kuwait in 1990, US and coalition forces assembled in Saudi Arabia, the biggest invasion force since World War II in a combined effort to evict Saddam Hussein's forces. The US took 1,848 Abrams tanks to the Gulf. The biggest transport aircraft operated by the US Air Force, the C-5 Galaxy, could carry only one tank at a time so most of them went by sea!

A massive air assault on Iraqi forces destroyed about half the Iraqi tanks and heavy armor before the ground war began, but the Abrams soon proved more than a match for anything the Iraqis had. Some feared the sophisticated Abrams, designed for a war in northern Europe, would fail in the hot desert and get bogged down in the shifting sand but this was not so.

▲

Racing into action an M1A1 prepares to fire on the move as a battle group surrounds enemy positions.

▲

Street fighting in Fallujah, Iraq. An Abrams returns fire from concealed anti-tank weapons in a house.

Abrams tanks spearheaded the ground attack and only 18 were taken out of action due to battle damage. With not one crewmember lost during the eviction of Iraqi forces from Kuwait in 1991, US commanders reported 90% operational readiness for their Abrams tanks, an unprecedented figure.

NEW ROLES FOR A NEW CENTURY

▲

A good view of skirt armor attached to lugs on the exterior of the main chassis. Individual plates can be changed in a matter of minutes.

In the fight to expel Iraqi forces from Kuwait in 1991, the Abrams MBT demonstrated its flexibility and outstanding performance. The Abrams was called to action when US forces invaded Iraq in 2003 and deposed Saddam Hussein. No tank was lost to hostile fire during the battle.

▲

Battle tanks are heavy and require strong bridges to cross rivers. Some Abrams have been converted to engineering support vehicles to throw bridges across water and dangerous abandoned trenches.

▲

Armor attached to the main structure of the tank is an adaption of the British Chobham armor, generally regarded as the best in the world.

However, on October 29, 2003, two soldiers were killed and a third wounded when their tank was disabled by an anti-tank mine. Just over a year later an improvised explosive device (**IED**) killed a third crewmember and destroyed the Abrams.

Designed during the Cold War and tested in Iraq, the M1A1 and M1A2 Abrams will continue to spearhead America's armed forces in any future ground war anywhere around the globe where US service members are called to action.

▲

Caught in winter snow an M1A1 uses a paved forest road during exercises in northern Germany.

Glossary

angular (ANG gyu lur) – something that has straight lines and sharp corners

armament (AR muh muhnt) – weapons and other equipment used for fighting wars

battalion (buh TAL yun) – a large unit of soldiers

blizzard (BLIZ urd) – a very heavy snowstorm

firepower (FIRE POU ur) – effective fire of weapons against a target

IED (Improvised Explosive Device) – unconventional weapons used by terrorist, most commonly roadside bombs

infrared (in FRA red) – outside the visible spectrum, used in night-vision equipment

MBT (Main Battle Tank) – a heavy tank that combines the ultimate in mobility, firepower, and protection

navigation (NAV uh gay shuhn) – the use of maps, compasses, GPS equipment to determine position and direction

periscope (PER uh skope) – a vertical tube with mirrors that allows you to see from a position below the tube

prototype (PROH tuh tipe) – the first version of an invention that test an idea to see if it will work

satellite (SAT uh lite) – a spacecraft in orbit around the earth often used in communication networks

terrain (tuh RAYNE) – the physical features of the land or ground

INDEX

FURTHER READING

Cornish, Geoff. *Tanks.* Lerner Publishing, 2003

Hunnicutt, R.P. Abrams. *A History of the American Main Battle Tank*, Volume 2. Presidio, 1990

Macksey, Kenneth. *Tank versus Tank*. Guild Publishing, 1988

WEBSITES TO VISIT

http://www.wikipedia.org/wiki/M1

http://www.army-technology.com/projects/abrams

ABOUT THE AUTHOR

David Baker is a specialist in defense and space programs, author of more than 60 books and consultant to many government and industry organizations. David is also a lecturer and policy analyst and regularly visits many countries around the world in the pursuit of his work.

**Indianapolis
Marion County
Public Library**

Renew by Phone
269-5222

Renew on the Web
www.imcpl.org

For General Library Information
please call 269-1700